THE
BIG
ENCYCLOPEDIA
OF DEFUNCT
ANIMALS

VOLUME VI

STANTON F. FINK

Acknowledgments

and Dedication

To my father, in whose books I discovered my first monsters.

To Nikolas Draper-Ivey, he who wields the sword of Ragnarok, and who has been awarded the Decorations of Omega; he who stands tall to cleave open the Heavens.

To Mariano Silvera, who should have had his own artbooks

To Doctor David Morafka, who helped teach me to be more picky with my information.

To my friends, who helped push me to make this.

Table of Contents

Introduction

The purpose of this coloring book series is to provide information on various prehistoric animals both profoundly famous and incredibly obscure to artists of all ages. Of course, there is a lot of material to work with, as animals have been a major component of Earth's ecosystems for at least 670 million years.

For the sake of space and workability, each volume will contain 17 entries: ideally, one species for each geological time period, if possible. If you, or your inner and or outer child do not see your favorite prehistoric animal here, it may be eventually featured in another volume. Or, contact me at apokryltaros@gmail.com to have it put into a later volume.

Glossary

- **Aquatic**- Living in water.
- **Arthropod**- Any member of the animal phylum Arthropoda, including trilobites, arachnids, crustaceans, insects, myriapods and their relatives. All arthropods have armor-like, jointed exoskeletons made of chitin-derived plates, sometimes reinforced with calcium carbonate, and jointed limbs.
- **Cambrian**- A period of time in the Paleozoic Era from 541 to 485 million years ago.
- **Carboniferous**- A period of time in the Paleozoic Era from 359 to 300 million years ago.
- **Cenozoic**- An era of time in the Phanerozoic Eon from 65 million years ago until now.
- **Chordate**- Any member of the animal phylum Chordata, including sea squirts, lancet fish, and vertebrates (such as lampreys, sharks, tuna, frogs, lizards, chickens, and people). All chordates have, at least at some point in their life cycle, a notochord, a long, flexible rod, usually made of cartilage, or, in the case of most vertebrates, cartilage and bone, running down the back from head to tail, directly beneath the neural tube.
- **Cnidarian**- Any member of the animal phylum Cnidaria, such as jellyfish, box jellies, Portuguese Man'o'war, sea anemones, coral and the parasitic myxozoans. Cnidarians are usually radially symmetrical, and have unique, venom-injecting stinging cells called "cnidocytes."
- **Cretaceous**- The last period of time in the Mesozoic Era, from 144 to 66 million years ago.
- **Devonian**- A period of time in the Paleozoic Era from 414 to 360 million years ago.
- **Ediacaran**- The last period of time in the Precambrian Eon from 635 to 542 million years ago.
- **Eocene**- A period of time in the Cenozoic Era from 55 to 33 million years ago.
- **Fauna**- In an ecological context, "fauna" refers to the animal components of an ecosystem.
- **Formation**- In a geological or paleontological context, a formation is a group of rock layers.
- **Gnathostome**- A gnathostome is any vertebrate chordate with a moveable jaw (or had an ancestor with one).
- **Holocene**- A period of time in the Cenozoic Era from 12,000 years ago until now.
- ***Incertae sedis***- A Latin phrase literally meaning "uncertain seat." *"Incertae sedis"* is a term in classification used to refer to a species or group whose relationships with related organisms are unclear or poorly defined.
- **Jurassic**- The second period of time in the Mesozoic Era, from 199 to 145 million years ago.
- **Mesozoic**- An era of time in the Phanerozoic Eon from 249 to 66 million years ago.
- **Miocene**- A period of time in the Cenozoic Era from 23 to 5 million years ago.

- **Mollusk**- Any member of the animal phylum Mollusca, including snails, clams, squid, octopuses, tusk shells and chitons. Most mollusks have a calcium carbonate shell, and a toothed, file-like tongue called a radula. All mollusks have a cape-like organ, the mantle, which usually secretes the shell, and houses breathing organs, and a nervous system.
- **Nekton**- Any aquatic animal that lives either entirely or almost entirely in the water column, and relies on its own swimming or propulsion abilities to keep and move itself in and around the water column. Anchovies, porpoises and ichthyosaurs are examples of nekton.
- **Neogene**- The second third of the Cenozoic Era, comprising of the Miocene and the Pliocene periods.
- **Oligocene**- A period of time in the Cenozoic Era from 33 to 23 million years ago.
- **Ordovician**- A period of time in the Paleozoic Era from 484 to 440 million years ago.
- **Paleocene**- A period of time in the Cenozoic Era from 65 to 55 million years ago.
- **Paleogene**- The first third of the Cenozoic Era, comprising of the Paleocene, Eocene, and Oligocene.
- **Paleozoic**- An era of time in the Phanerozoic Eon from 249 to 66 million years ago.
- **Permian**- The last period of time in the Paleozoic Era, the time of "The Great Dying," or most severe of all known extinction events, from 299 to 250 million years ago.
- **Pharynx**- A structure in the throat of many animals located directly behind the mouth or oral chamber. In vertebrates, it often houses breathing structures, like gills.
- **Plankton**- An organism that uses water currents and waterflow to as its primary means of transportation in the water column because it is either too small to move long distances by its own power, or lacks the ability to propel itself entirely. Sargassum seaweed and jellyfish are two varieties of plankton.
- **Pleistocene**- A period of time in the Cenozoic Era from 3 million years ago until 12 thousand years ago.
- **Pliocene**- A period of time in the Cenozoic Era from 5 to 3 million years ago.
- **Quaternary**- The last third of the Cenozoic Era, comprising of the Pleistocene and the Holocene periods.
- **Terrestrial**- Living on land.
- **Triassic**- The first period of time in the Mesozoic Era, from 249 to 200 million years ago.

Name	*Fedomia mikhaili*
Phylum	Porifera
Class	*incertae sedis*
Size	Arms ranging from 1 to 5 centimeters in length, diameter about 2 to 5 millimeters.
Time Period	Late Ediacaran, 558 to 555 million years ago
Location	Solza River, near the White Sea, Russia
Comments	*Fedomia mikhaili* is a sponge-like Precambrian fossil that has several hollow, sausage-shaped branches whose bases connect at the top of a low, pedestal-like stalk. The outer surfaces of the branches were covered in clusters of spines, probably derived from sponge spicules, giving the living organism a superficial resemblance to a modern-day cactus plant. If *F. mikhaili* was a sponge, then it would have been a filter-feeder that drew water through its porous bodywalls in order to strain out plankton and other edible particles from the water.

Name *Kodymirus vagans*

Phylum Arthropoda

Subphylum Chelicerata

Class ?Aglaspidida

Order *incertae sedis*

Size The living animal is estimated to be around 8 centimeters long.

Time Period "Series 2" of the Early Cambrian, about 515 million years ago.

Location Paseky Shale, Czech Republic.

Comments *Kodymirus vagans* is a very large cheliceratan arthropod from the Early Cambrian of the Czech Republic. Its fragmentary remains come from the Paseky Shale, which, during the Early Cambrian, was the earliest known example of an estuarine community.

The large forelimbs are thought to have assisted in both swimming and seizing prey. *K. vagans* may be related to the mostly Early Paleozoic arthropod group Aglaspidida, though more better-preserved fossils are needed to confirm this hypothesis.

Name	*Cryptocrinites laevis*
Phylum	Echinodermata
Class	Eocrinoidea
Family	Cryptocrinitidae
Size	Theca (body) about 1 centimeter in diameter.
Time Period	Aseri Stage of the Darriwllian Epoch, Middle Ordovician period, 463 to 461 million years ago.
Location	Estonia, Norway, Saint Petersburg area of Russia.
Comments	*Cryptocrinites laevis* is an extinct eocrinoid echinoderm that lived in shallow, probably murky seas in what is now Baltic Russia, Estonia and Norway.

Most fossils of *C. laevis* are of its theca, which is similar in size and shape to a garbanzo bean. The plates that make up its theca are arranged in a spiral, which can be observed by looking at the top. Like other eocrinoids, *C. laevis* was a filter-feeder that lived attached to the seafloor by a holdfast. The arm-like brachioles at the top of the theca housed tiny tube-feet that captured edible particles and transferred them to its mouth. The eye-like structure is the madreporite, which pumped water into the animal's hydrovascular system.

Name	*Spirocrinus uniformis*
Phylum	Echinodermata
Class	Crinoidea
Family	Petalocrinidae
Size	Arm 2 to 3 centimeters long, crown probably up to 8 centimeters in diameter.
Time Period	Llandovery Epoch, Silurian period, 444 to 428 million years ago.
Location	Leijiatun Formation, Guizhou Province, China.
Comments	*Spirocrinus uniformis* is a species of petalocrinid sea lily, or crinoid echinoderm that lived on the ocean floor in what is now Guizhou Province, during the Silurian Period. Petalocrinids are unusual among sea lilies because their arms are formed from all of the tiny units of the arm fusing into a solid wedge shaped, depending on the genus, like a guitar pick, a button, a flower petal, or, in the case of *Spirocrinus*, a cigar. The arms of petalocrinids are assumed to have been able to capture food, as implied by the deep grooves in the oral (upper) surfaces of the arms.

Because China has three endemic genera (the Ordovician *Eopetalocrinus*, and the Silurian *Spirocrinus*, and *Sinopetalocrinus*), and the highest number of endemic species, it is assumed that Petalocrinidae originated in China. Much like the guitar pick-shaped "arms" of *Petalocrinus*, the cigar-shaped arms of *Spirocrinus* helped keep the body of the animal up in the water colum while allowing it to filter-feed.

Name	*Loriolaster mirabilis*
Phylum	Echinodermata
Class	Ophiuroidea
Order	Lysophiurina
Family	Encrinasteridae
Size	Body and arm up to about 10 centimeters in diameter.
Time Period	Emsian epoch of the Early Devonian period, 407 to 393 million years ago
Location	Gemünden municipality, Rhein-Hunsrück, Germany
Comments	*Loriolaster mirabilis* is an unusual brittlestar in the unusual, Devonian brittlestar family Encrinasteridae. Encrinasterid brittlestars had unusually broad, dinner plate-like bodydiscs (the normally small, often pentagon-shaped portion of the body where the arms are attached, and which houses the stomach). In the case of the Hunsrück *Loriolaster*, the bodydisk was stretched thin, like a tent's canvas, to the tips of the arms. Its arms, in turn, were embedded in the skin of the bodydisk, with their numerous spines sticking out of the oral (bottom) side. What sort of function this tent-like bodydisc performed remains unknown.

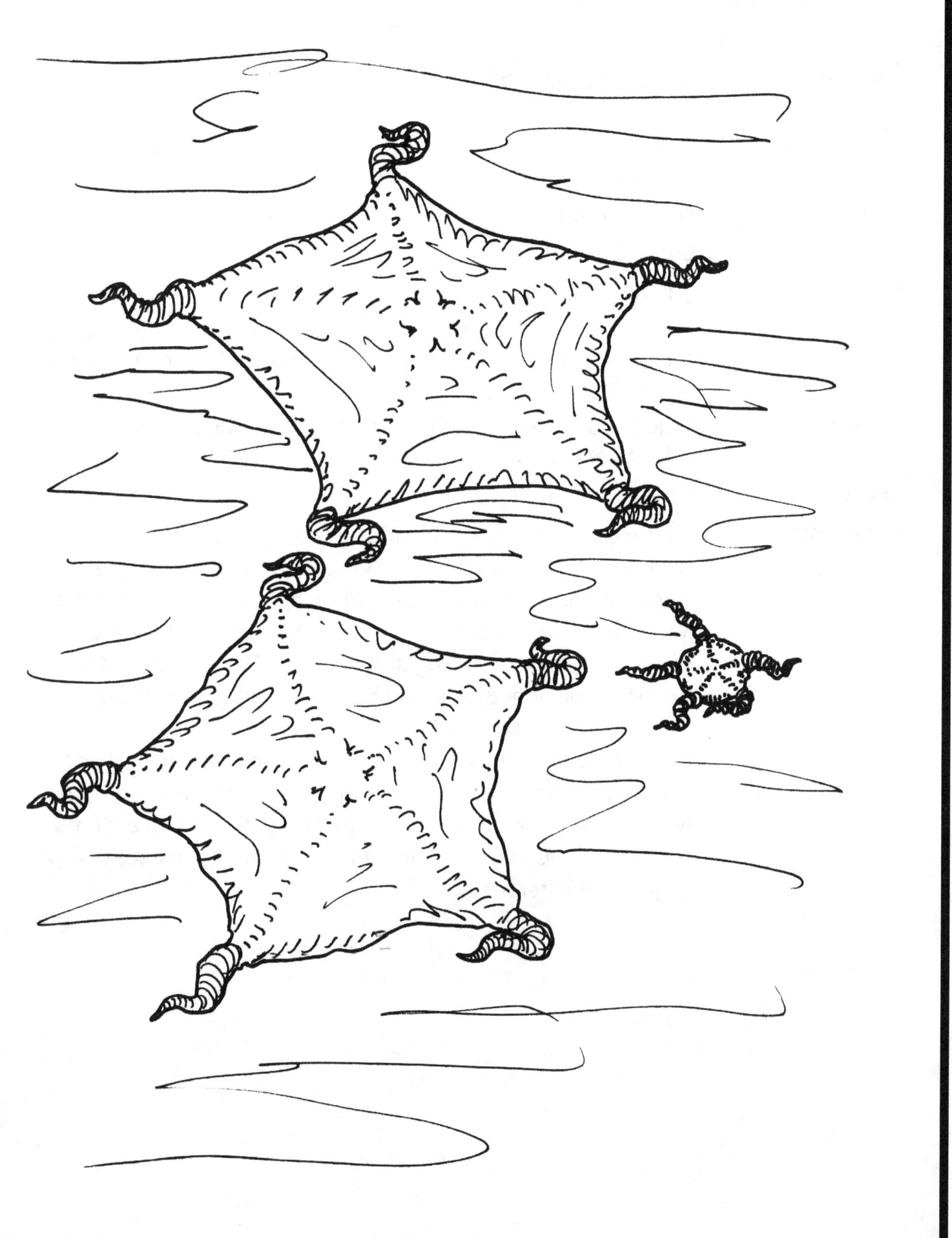

Name	*Gastrioceras listeri*
Phylum	Mollusca
Class	Cephalopoda
Subclass	Ammonoidea
Order	Goniatitida
Family	Gastrioceratidae
Size	Adult shells generally 4 centimeters in diameter, some exceptional individuals may have shells up to 9 centimeters in diameter.
Time Period	Langsettian Stage of the Late Carboniferous, 313 to 311 million years ago.
Location	Europe, China, Algeria
Comments	*Gastrioceras listeri* is a widespread species of goniatitid ammonoid that lived in shallow seas during the Langsettian Stage of the Carboniferous (other species of *Gastrioceras* would persist until the Late Permian). The first fossils of *G. listeri* were found in England in the 19th Century. Other locations include Belgium, The Netherlands, Germany, the Ukraine, Algeria and China. Because *G. listeri* has a broad, almost globular shaped shell, it probably was a benthic-dwelling ammonoid that either swam very close to, or crawling along the surface of the seafloor. *G. listeri* may have fed upon small crustaceans or large foraminiferans.

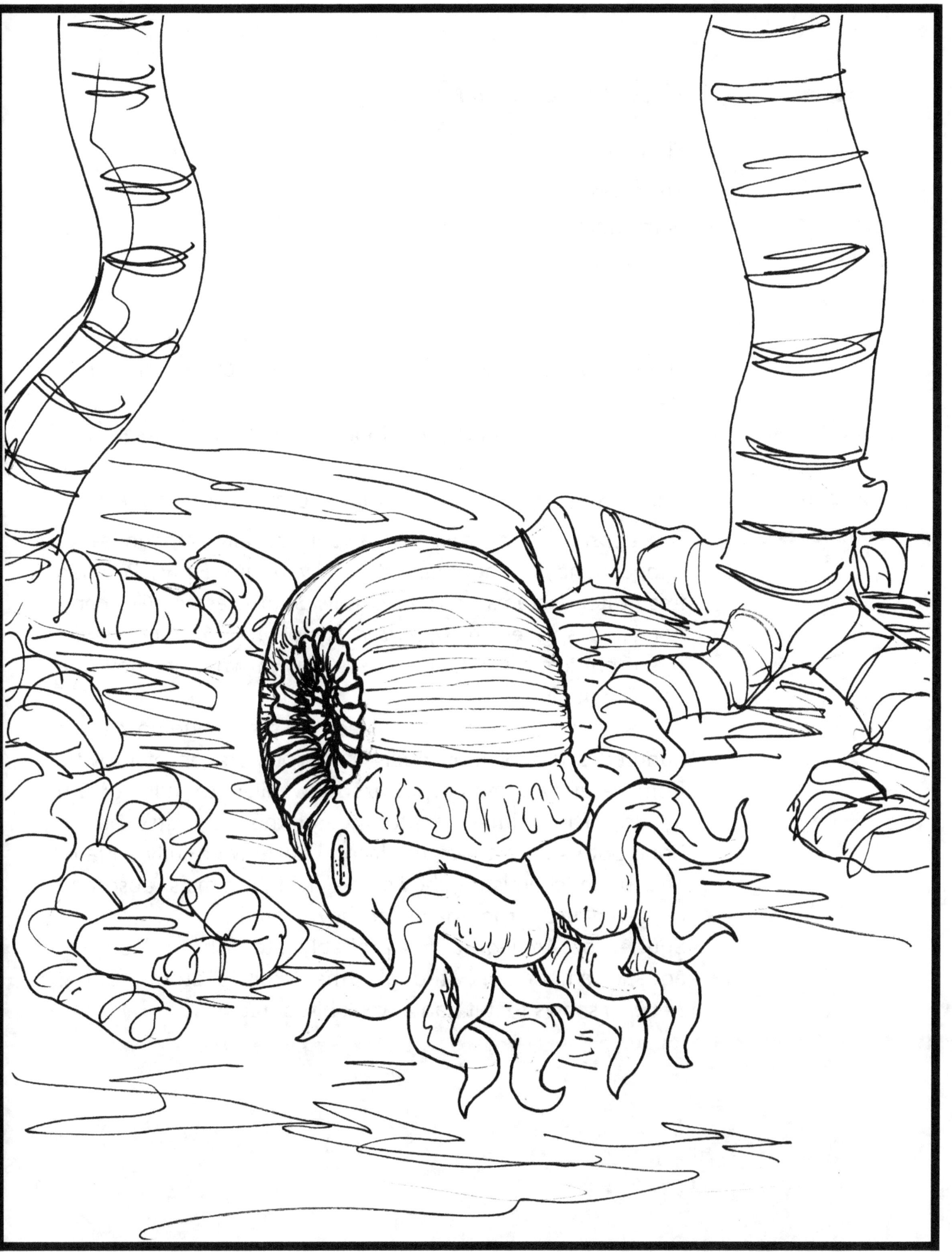

Name	*Archegosaurus decheni*
Phylum	Chordata
Class	Amphibia
clade	Batrachomorpha
Order	Temnospondyli
Family	Archegosauridae
Size	Skulls ranging in size from 2 centimeters (juveniles) to almost 30 centimeters (adults), adults probably similar in size to a dwarf caiman
Time Period	Assellian to Wuchiapingian Stages of the Early Permian, 299 to 253 million years ago.
Location	Lake Humberg, Saar-Nahe Basin, Southwestern Germany
Comments	*Archegosaurus decheni* is a superficially crocodile-like temnospondyl amphibian that lived in Early Permian freshwater systems in what is now Southwestern Germany. A related species, *A. dyscriton* (synonym *Memonomemos dyscriton*) lived in freshwater systems in what are now the Ural Mountains.

A. decheni is known from several fossils, including over a hundred skulls of individuals in various stages of growth, from recently metamorphosized juveniles to fully grown adults. The animal's anatomy, and the chemistry of the bone both suggest that the living beasts were fully aquatic. The bone chemistry shows the living animals absorbed oxygen directly from the water they swam in.

Species of *Archegosaurus* probably ate fish, and smaller aquatic amphibians, though, the skull anatomy of *A. decheni* shows that it had a weak bite force.

Name	*Ticinepomis peyeri*
Phylum	Chordata
Class	Sarcopterygii
Order	Coelacanthiformes
Family	Latimeriidae
Size	Up to 18 centimeters long
Time Period	Lower Ladinian Epoch of the Middle Triassic, 491 million years ago
Location	Canton Ticino, Monte San Giorgio, Switzerland
Comments	*Ticinepomis peyeri* is an extinct coelacanth from marine strata in what is now the mountains of southern Switzerland along the Italian border. *T. peyeri* is a relative of both the modern coelacanth, *Latimeria sp.*, and the bizarre *Foreyia maxkuhi,* which also lived with *Ticinepomis* in Canton Ticino. The living animal would have a strong resemblance to the modern coelacanth (as do most Mesozoic coelacanths), differring in having a more elongated snout, and more delicate fins.

Detailed examinations of the fossils of *Foreyia* would reveal that it and *Ticinepomis* were closely related, sharing numerous anatomical features despite their dramatically different forms. |

Name	*Phlycticeras buckmani*
Phylum	Mollusca
Class	Cephalopoda
Subclass	Ammonoidea
Order	Ammonitida
Family	Strigoceratidae
Size	Adult macroconch (female's shell) up to 8 centimeters in diameter, 2 centimeters wide. Adult microconch (male's shell) up to 2 centimeters long.
Time Period	Late Bajocian Epoch of the Middle Jurassic Period, about 168 million years ago.
Location	Ferruginous oolite of *Parkinsoni* Zone, near Bayeux, Normandy, France.
Comments	*Phlycticeras buckmani* is a sexually dimorphic ammonite from the Middle Jurassic of what is now Normandy, France. What is assumed to be the macroconch (literally "big shell") or female is a robust cookie-sized shell that was originally described as *Phlycticeras buckmani*. The macroconch has numerous, very pronounced ribs. What is assumed to be the microconch (i.e., "small shell), or male is a bean-sized and urn-shaped shell originally described as *Oecoptychius grossouvrei*. Because *Oecoptychius grossouvrei* is also found in the same localites as *P. buckmani*, and is often found in association with fossils of *P. buckmani*, *O. grossouvrei* is assumed to be the male of the species (given as how the vast majority of invertebrate sexual dimorphism involves tiny males paired with enormous females). Most species of *Oecoptychius* are now assumed to be the microconchs of several species of *Phlycticeras*, though, in the case of *P. polygonium*, the species *P. schaumburgi* is thought to be the former's similarly-sized microconch.

Name *Dsungaripterus weii*

Phylum   Chordata

clade   Archosauria

clade   Pterosauromorpha

Order   Pterosauria

Suborder   Pterodactyloidea

Family   Dsungaripteridae

Size   Wingspan 3 to 3.5 meters, skull length from 40 to 50 centimeters.

Time Period Tsaganstabian Age of the Early Cretaceous, about 132 to 129 million years ago.

Location   Dzungarian (or Junggar) Basin, Dzungaria, Northern Xinjiang Province, China.

Comments   *Dsungaripterus weii* is an unusual pterodactyloid pterosaur from the Early Cretaceous of Western China. It had an elongated, low-crested skull with the jaws modified into a stout, upward-pointing, tweezer-like form with flattened, knob-like teeth in the back. This highly modified skull, coupled with its stout (for a pterosaur) skeleton, suggest to researchers that the living animal flew from mudflat to mudflat to wade around in mud, and spear burrowing shellfish, who, upon capture, would be thrown into the back of the beast's mouth to be subdued with an armor-cracking crunch.

Here, an adult *D. weii* is compared to its smaller relative, *Noripterus parvus*.

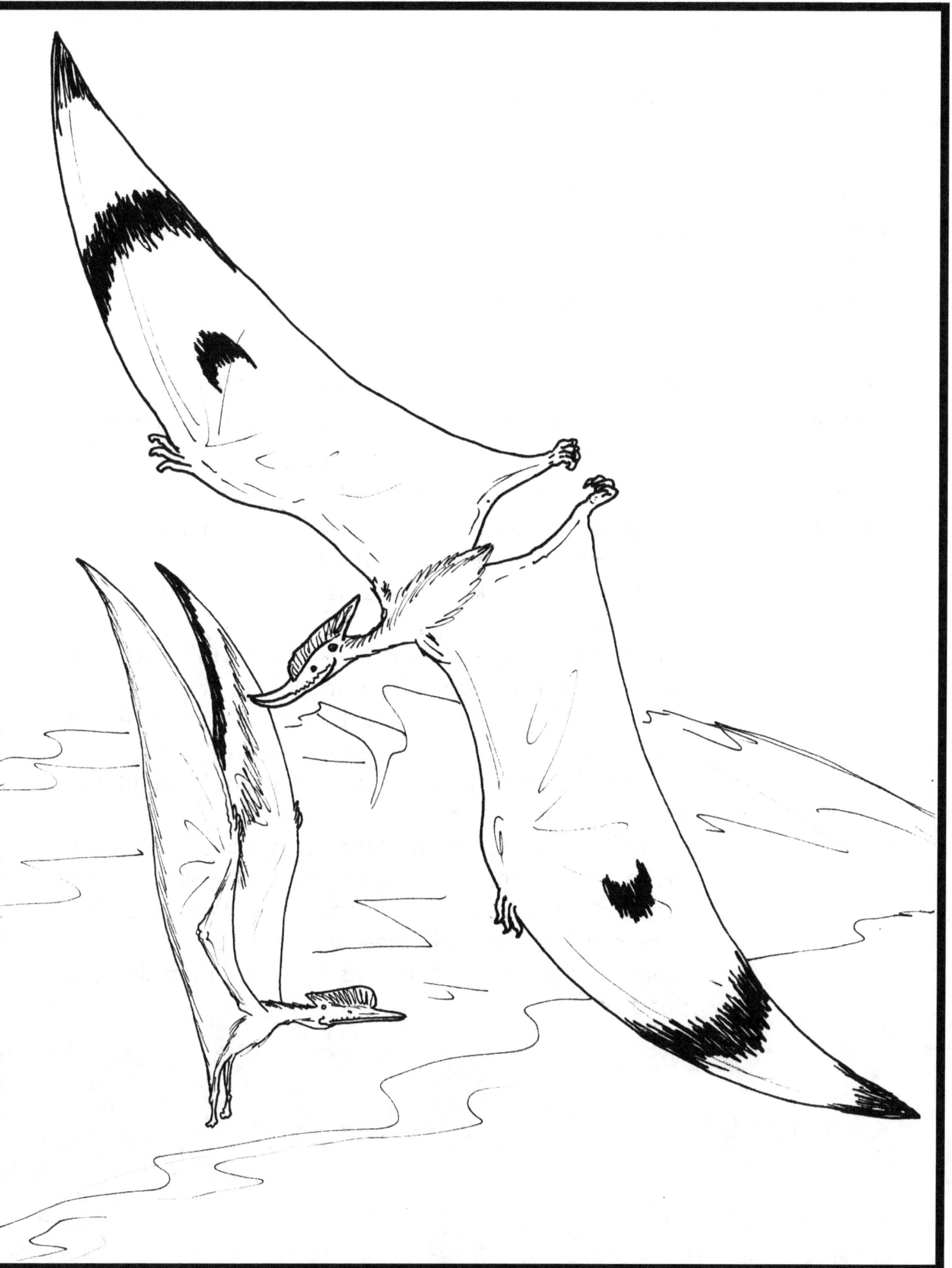

Name   *Corydoras revelatus*

Phylum   Chordata

Class   Actinopterygii

Order   Siluriformes

Family   Callichthyidae

Subfamily   Corydoradinae

Tribe   Corydoradini

Size   Holotype specimen about 3 centimeters long.

Time Period   Late Thanetian Stage of the Late Paleocene Period, 58 million years ago

Location   Mais Gordo Formation, Salta, Argentina.

Comments   *Corydoras revelatus* is an extinct species of armored catfish in the family Callichthyidae. The holotype and only known specimen is from late Paleocene strata of Salta Province, Argentina.

The fossil is very similar in anatomy to modern-day *Corydoras* species, but differs in that *C. revelatus* has a shorter, more rounded or blunt head, and lower-set eyes. Even so, *C. revelatus* undoubtedly had a lifestyle identical to modern species, grubbing through sediment for detritus and burrowing invertebrates.

Experts remain unsure which modern species *C. revelatus* is most closely related to. Experts do understand, however, that its modern appearance means that callichthyid catfishes were already diversifying either soon after or before the KT Extinction Event 66 million years ago (and 12 million years prior).

Name	*Harpagolestes macrocephalus*
Phylum	Chordata
Class	Mammalia
Order	Mesonychia
Family	Mesonychidae
Size	Partial skull, humerus and other ragmentary remains imply a living animal similar in size to a black bear.
Time Period	Wasatchian land stage of the Early Eocene, 55 to 50 million years ago
Location	"...lower part of Bridger beds near the mouth of Smith's Fork", in the vicinity of Millersville, Wyoming.
Comments	*Harpagolestes macrocephalus* is the type species of the mesonychid genus *Harpagolestes*. Species of *Harpagolestes* are found in Early to Middle Eocene strata of China and North America, and are closely related to the mesonychid genus *Mesonyx,* species of which share very similar dental anatomy. Legbones and other scrappy postcranial remains suggest that *H. macrocephalus* and other species of *Harpagolestes* had much stouter, more robust builds than the "wolf" like builds of *Mesonyx* species. These, and the robust skulls suggest that *H. macrocephalus* and other *Harpagolestes* species probably scavenged or stole the kills of smaller predators, and that the wear on their often very worn-down teeth show that they crunched down on bones for much of their diet.

Name	*Eoanabas thibetana*
Phylum	Chordata
Class	Actinopterygii
Order	Anabantiformes
Family	Anabantidae
Size	Average bodylength about 3 centimeters.
Time Period	Chattian stage of the Late Oligocene, 26 to 23.5 million years ago.
Location	Holotype specimen found in Jiangnongtangga of the upper portion of the Dingqing Formation, in the southern Nima Basin, Central Tibet. Other specimens found in Songwori, also in the Nima Basin, and in Dayu, of the Lunpola Basin.
Comments	*Eoanabas thibetana* is a small, extinct climbing gourami or climbing perch that lived in a lake system in what is now Central Tibet, soon after India collided with Asia, but just before or at the beginning of the uplift that would raise the Tibetan Plateau.

E. thibetana is an important fossil for several reasons: one being that it is the "stereotypical ancestral member" that possesses features otherwise unique to certain descendants, and the other being that it gives clues about the palaeoenvironment it lived in.

E. thibetana suggests an animal that, when alive, looked very much like modern Asian climbing gouramis (i.e., of genus *Anabas*), but, also possessed features seen only in African climbing gouramis (in that fossils identified as male possess a spine otherwise only seen in *Microctenopoma* species used to stimulate the females into releasing eggs during mating).

A fossil climbing perch, coupled with fossil leaves of tropical trees now living in Southeastern China strongly suggest that, during the late Oligocene, Tibet was covered in wet, tropical forests interwoven with numerous lake systems.

Name   *Necronectes proavitus*

Phylum	Arthropoda
Subphylum	Crustacea
Class	Malacostra
Order	Decapoda
Infraorder	Brachyura
Family	Portunidae
Subfamily	Necronectinae
Size	Carapace average length 4.5 centimeters, carapace average width 7.5 centimeters
Time Period	Middle to Late Miocene
Location	Caribbean coasts of Mexico, Central America and northern South America.

Comments

Necronectes proavitus is an extinct species of swimming crab related to the still-extant mud and mangrove crabs of the Asian genus *Scylla*. The genus *Necronectes* probably originated along the Pacific coast of Mexico, as the earliest fossils are known from middle Eocene Baja California. During the Oligocene, the genus spread into the Caribbean, the southern United States, and across the Atlantic to Iberia, where fossils of the type species *N. vidalianus* are found in French Oligocene-aged marine strata.

N. proavitus, which has a more circular carapace compared to other species, is found in in Middle to Late Miocene marine strata along the Mexican to South American coasts of the Caribbean Sea. It is the youngest known species, and probably died out due to climate change brought about by the separation of the Pacific and Atlantic Oceans because of the formation of the Isthmus of Panama.

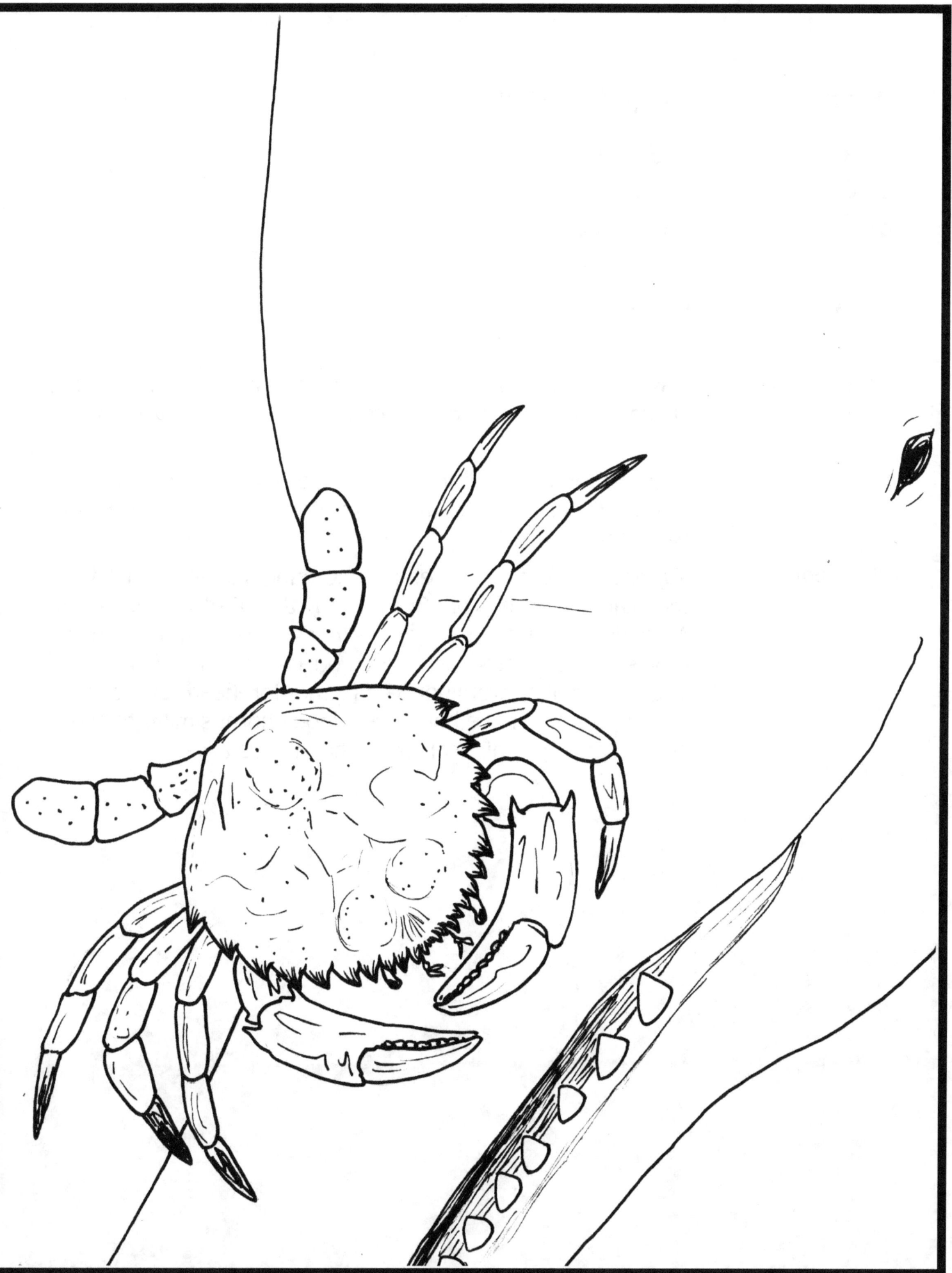

Name	*Trigodon gaudryi*
Phylum	Chordata
Class	Mammalia
Order	Notoungulata
Family	Toxodontidae
Subfamily	Haplodontitheriinae
Size	Skull about 54 centimeters long.
Time Period	Mayoan stage until the Montehermosan stage, from Late Miocene until the Early Pliocene, 11.6 million to 4 million years ago.
Location	Miocene fossils found in Solimões Formation, Acre Province, Brazil. Pliocene fossils found in Montehermoso Formation, Patagonia, Argentina
Comments	*Trigodon gaudryi* is an extinct notoungulate mammal from the Late Miocene of Brazil and the Early Pliocene of Argentina. It was a large, heavy-set animal that probably browsed and grazed on grass and low-growing foliage. A large knob-like prominence on the forehead of its skull suggests it may have had a keratinous horn, similar to those of the unrelated rhinoceroses, when it was alive.

Name	*Pyrazisinus ultimus*
Phylum	Mollusca
Class	Gastropoda
taxon	Caenogastropoda
Superfamily	Cerithioidea
Family	Potamididae
Size	Holotype specimen about 53 millimeters, adult shell lengths ranging from 48 to 55 millimeters
Time Period	Late Sangamonian stage of the Late Pleistocene, about 75,000 years ago.
Location	Coffee Hill Hammock Member of the Fort Thompson Formation in the Okeechobee Group, near Wellington, Palm Beach County, Florida.
Comments	*Pyrazisinus ultimus* is an extinct species of horn snail or mudwhelk from Late Pleistocene Florida. It is a typical-looking member of the family Potamididae, having a long, spire-shaped shell with ribs. In life, *P. ultimus*, much like other mudwhelks, lived in mudflats associated with mangrove forests, and probably ate algae and biofilms. Edward Petuch gave this species the epithet *"ultimus"* meaning "the last" in reference to how this appears to be the last species in the genus *Pyrazisinus* before going extinct in the Late Pleistocene. *P. ultimus* probably went extinct when the mangrove forest it lived in dried out due to changing sea levels effectively moving the location away from the Atlantic Ocean.

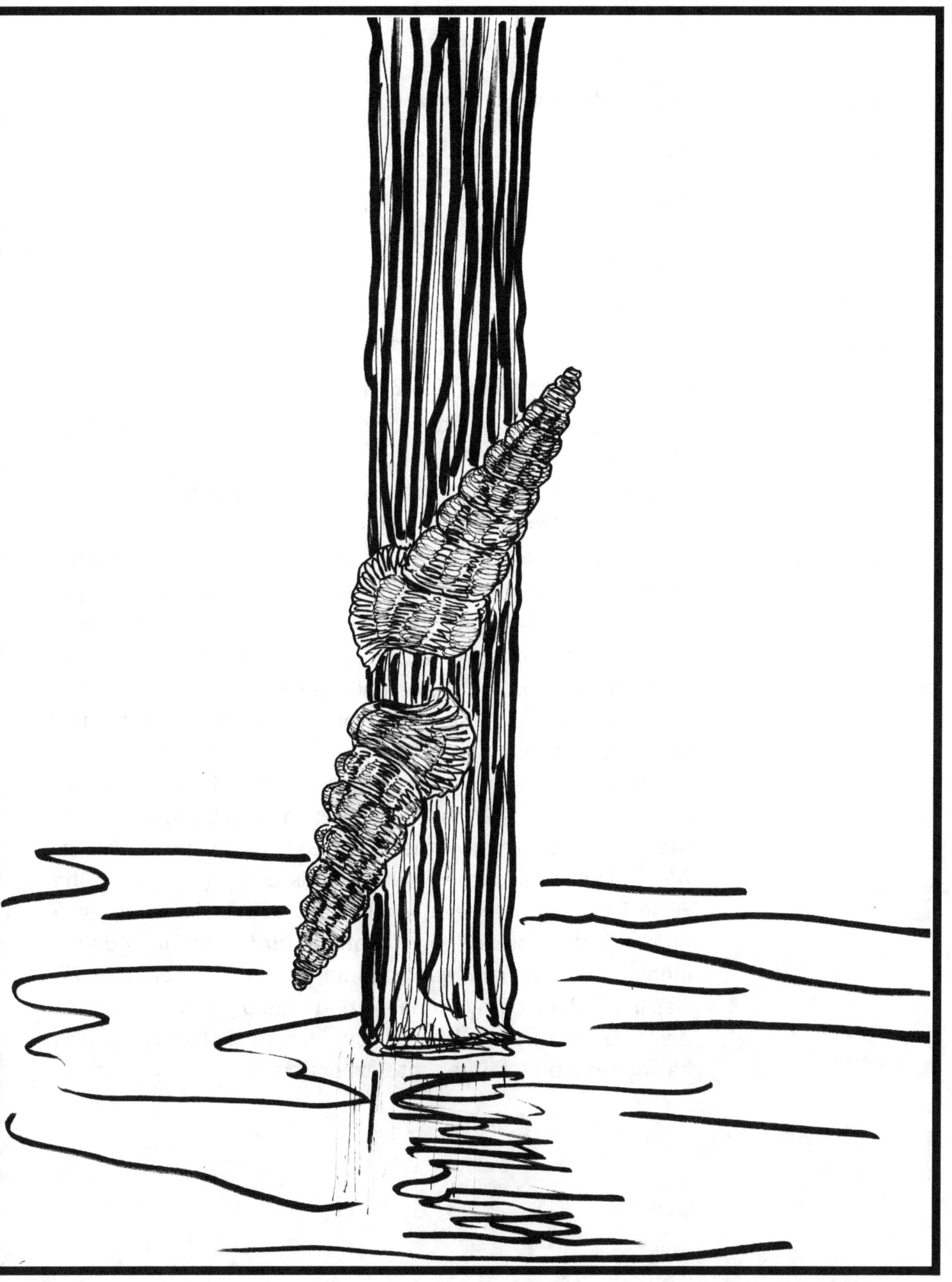

Name	*Volia athollandersoni*
Phylum	Chordata
Class	Reptilia
clade	Archosauria
Order	Crocodilia
Family	Crocodylidae
Subfamily	Mekosuchinae
Size	Estimated to be about 2 to 3 meters long.
Time Period	Late Pleistocene to Late Holocene, about 13,000 to 3,000 years ago (~1000 BCE)
Location	Subfossils found in Voli-Voli Cave and Wainibuku Cave from Wainibuku Valley, Viti Levu, Fiji.
Comments	*Volia athollandersoni* is an extinct mekosuchine crocodile from the rainforests of ancient Fiji. *V. Athollandersoni* is one of the last species and the youngest species of the crocodile subfamily Mekosuchinae, a group that was otherwise originally endemic to Australia since the Eocene. This diverse group would slowly die out in Australia, until the last few species there went extinct during the late Pleistocene. The long-lived genus *Mekosuchus* emigrated into Polynesia, first by traveling to the now sunken Greater Chesterfield Island in what is now the Coral Sea. The individuals of *Mekosuchus* that came to the islands of Fiji (or at least the Fijian island of Viti Levu) would evolve into *Volia*, an apex predator that preyed on flightless birds and indigenous iguanas. *V. Athollandersoni* would go extinct soon after Fiji was first colonized by humans about 3,000 years ago, due to being both directly hunted by ancient Polynesians, and having their prey hunted into extinction.

Bibliography

- Aguilera, Orangel A., and Alfredo A. Carlini, eds. *Urumaco and Venezuelan paleontology: The fossil record of the Northern Neotropics*. Indiana University Press, 2010.
- Arkell, W. J., W. M. Furnish, and Bernhard Kummel. *Treatise on Invertebrate Paleontology, Part L: Mollusca 4, Cephalopoda, Ammonoidea*. Geological Society of America, 1957.
- Agustí, Jordi, Mauricio Antón, and Mauricio Anton. *Mammoths, sabertooths, and hominids: 65 million years of mammalian evolution in Europe*. Columbia University Press, 2002.
- Bartels, Christoph, Derek EG Briggs, and Günther Brassel. *The fossils of the Hunsrück Slate: marine life in the Devonian*. Cambridge University Press, 1998.
- Bennett, S. Christopher. "New information on body size and cranial display structures of *Pterodactylus antiquus*, with a revision of the genus." *Paläontologische Zeitschrift* 87.2 (2013): 269-289.
- Blieck, Alain. "Les Hétérostracés Ptéraspidiformes, Agnathes du Silurien-Dévonien du Continent nord-atlantique et des blocs avoisinants: révision systématique, phylogénie, biostratigraphie, biogéographie." (1984).
- Bockelie, J. FREDRIK. "The Middle Ordovician of the Oslo region, Norway, 30. The eocrinoid genera Cryptocrinites, Rhipidocystis and Bockia." *Norsk Geologisk Tidsskrift* 61 (1981): 123-147.
- Cavin, Lionel, et al. "Heterochronic evolution explains novel body shape in a Triassic coelacanth from Switzerland." *Scientific Reports* 7.1 (2017): 13695.
- Chlupáč, I., and V. Havlíček. "Kodymirus ng, a new aglaspid merostome of the Cambrian of Bohemia." *Sborník Geologickych Věd. Paleontologie* 6 (1965): 7-20.
- Fedonkin, Mikhail A., et al. *The rise of animals: evolution and diversification of the kingdom Animalia*. JHU Press, 2007.
- Frickhinger, Karl Albert. *Fossil atlas, fishes*. Mergus, 1995.
- Lamsdell, James C., Martin Stein, and Paul A. Selden. "Kodymirus and the case for convergence of raptorial appendages in Cambrian arthropods." *Naturwissenschaften* 100.9 (2013): 811-825.
- Harrington, H. J. "General description of Trilobita." *Treatise on invertebrate paleontology, Part O, Arthropoda* 1 (1959): 38-117.
- Hone, D. W. E., S. Jiang, and X. Xu. "A taxonomic revision of Noripterus complicidens and Asian members of the Dsungaripteridae." *Geological Society, London, Special Publications* 455.1 (2018): 149-157.
- Lundberg, John G., et al. "Discovery of African roots for the Mesoamerican Chiapas catfish, Lacantunia enigmatica, requires an ancient intercontinental passage." *Proceedings of the Academy of Natural Sciences of Philadelphia* 156.1 (2007): 39-54.
- Mao, Ying-Yan, et al. "Chinese origin and radiation of the Palaeozoic crinoid family Petalocrinidae." *Palaeoworld* 24.4 (2015): 445-453.
- Molnar, R. E., T. Worthy, and P. M. A. Willis. "An extinct Pleistocene endemic mekosuchine crocodylian from Fiji." *Journal of Vertebrate Paleontology* 22.3 (2002): 612-628.
- Novitskaya, L. I. "Evolution of generic and species diversity in agnathans (Heterostraci: Orders Cyathaspidiformes, Pteraspidiformes)." *Paleontological Journal* 41.3 (2007): 268-280.
- Petuch, Edward J. *Cenozoic seas: the view from eastern North America*. CRC Press, 2003.
- Prothero, Donald R., and Scott E. Foss, eds. *The evolution of artiodactyls*. JHU Press, 2007.
- Rieppel, O. "A new coelacanth from the Middle Triassic of Monte San Giorgio,

Switzerland." *Eclogae Geologicae Helvetiae* 73.3 (1980): 921-939.

- Schweigert, Günter, and Volker Dietze. *Revision der dimorphen Ammonitengattungen Phlycticeras Hyatt-Oecoptychius Neumayr (Strigoceratidae, Mitteljura)*. Staatl. Museum für Naturkunde, 1998.

- Schweitzer, Carrie E., et al. "New crabs from the Eocene and Oligocene of Baja California Sur, Mexico and an assessment of the evolutionary and paleobiogeographic implications of Mexican fossil decapods." *Journal of Paleontology* 76.sp59 (2002): 1-44.

- Solé, Floréal, et al. "The European Mesonychid Mammals: Phylogeny, Ecology, Biogeography, and Biochronology." *Journal of Mammalian Evolution* 25.3 (2018): 339-379.

- Szalay, Frederick S., and Stephen Jay Gould. "Asiatic Mesonychidae (Mammalia, Condylarthra). Bulletin of the AMNH; v. 132, article 2." (1966).

- Turner, Alan. *National Geographic Prehistoric Mammals*. National Geographic, 2004.

- Witzmann, Florian, and Rainer R. Schoch. "The postcranium of Archegosaurus decheni, and a phylogenetic analysis of temnospondyl postcrania." *Palaeontology* 49.6 (2006): 1211-1235.

- Wu, Feixiang, et al. "Fossil climbing perch and associated plant megafossils indicate a warm and wet central Tibet during the late Oligocene." *Scientific reports* 7.1 (2017): 878.

About the Artist

Stanton F. Fink is a student of Biology and Chinese Medicine, and makes a hobby of drawing monsters and researching flowers, arcane-looking creatures, prehistoric animals, fish, reptiles, birds and the occasional, really grotesque fungal fruiting body.

Stanton grew up and went to school in California and is currently living, drawing, and gardening in Oregon.